EXTREME BIOLOGY

Eat or Be Eaten

Extreme Food Chains

Louise Spilsbury

Gareth Stevens
PUBLISHING

Please visit our website, **www.garethstevens.com**. For a free color catalog of all our high-quality books, call toll free 1-800-542-2595 or fax 1-877-542-2596.

Library of Congress Cataloging-in-Publication Data

Spilsbury, Louise, author.
 Eat or be eaten : extreme food chains / Louise Spilsbury.
      pages cm. — (Extreme biology)
 Includes bibliographical references and index.
ISBN 978-1-4824-2239-9 (pbk.)
ISBN 978-1-4824-2240-5 (6 pack)
ISBN 978-1-4824-2238-2 (library binding)
1. Food chains (Ecology)—Juvenile literature. 2. Ecology—Juvenile literature. I. Title.
QH541.14.S677 2015
577.16—dc23

2014027565

First Edition

Published in 2015 by
Gareth Stevens Publishing
111 East 14th Street, Suite 349
New York, NY 10003

© 2015 Gareth Stevens Publishing

Produced by: Calcium, www.calciumcreative.co.uk
Designed by: Paul Myerscough
Edited by: Sarah Eason and John Andrews
Picture research by: Rachel Blount

Photo credits: Cover: Shutterstock: Bruno Laveissiere; Inside: Ardea: Yves Bilat 26; Dreamstime: Steve Allen 44, Stef Bennett 34, Lukas Blazek 23, Chriswood44 41, Cupertino10 3, 8, Catherine Downie 32, Edurivero 33, Risto Hunt 30, Sylvie Lebchek 24, Mietitore 31, Outdoorsman 29, Somyot Pattana 25, Peternile 1, 10, Bhalchandra Pujari 42, Morley Read 12, Chris Sargent 28, Vladimir Seliverstov 40, Teguh Tirtaputra 36; Shutterstock: Ajman 45, Greg Amptman 7, Amskad 35, Petrov Andrey 20, Anton Ivanov 17, Eniko Balogh 18, Hagit Berkovich 22, BMJ 38, Ethan Daniels 6, Tyler Fox 19, Tami Freed 14, Geo-Zlat 21, Steven Gill 39, Jody. 13, Johnlips 4, Kjersti Joergensen 15, Micha Klootwijk 27, Longjourneys 9, MarcusVDT 16, Juriah Mosin 37, Konstantin Novikov 5, Sergey Uryadnikov 11, Monika Wieland 43.

Printed in the United States of America
CPSIA compliance information: Batch #CW15GS: For further information contact Gareth Stevens, New York, New York at 1-800-542-2595.

Contents

Chapter 1

Ocean Warfare

The surface of an ocean often looks calm and quiet. However, beneath the surface, it is a frantic, dangerous place where animals hunt and eat each other in the battle for survival. All living things need energy to stay alive, and most get energy by eating other living things. The series of living things that eat each other is called a food chain.

Plankton Producers

Many ocean food chains begin with tiny creatures called plankton. Plankton includes floating plantlike phytoplankton and weakly swimming animals that live near the surface of the ocean. Most are so small they can be seen only through a microscope. Phytoplankton organisms do not feed on other things. They use energy from the sun to make food inside their bodies from water and air. This process is called photosynthesis. Living things that make their own food are called producers, because they produce food for other organisms.

Tiny floating animals like this that are found in plankton are known as zooplankton.

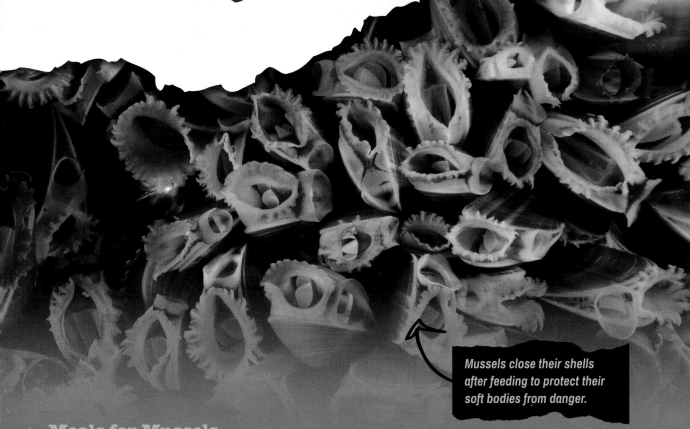

Mussels close their shells after feeding to protect their soft bodies from danger.

Meals for Mussels

Many different animals feed on plankton, including mussels. Mussels are a type of shellfish. They attach themselves to rocks and wait for the ocean to wash floating plankton toward them. To eat, a mussel pokes two tubes out of its shell. It sucks water in through one tube and passes it through some slimy threads inside its shell. These threads filter out the plankton for the mussel to eat. The mussel then spits the water out through the other tube.

Extreme!

Feeding an Ocean

Phytoplankton organisms are so small that there can be 1 million in just a teaspoon of water. However, they are the most important and abundant creatures in ocean food chains. Animal plankton, fish, and larger ocean animals such as whales all depend on plankton for their survival.

Spiny Lobster

The alien-looking spiny lobster is a formidable ocean predator. It has an armored shell with strong, forward-facing spines and can grow up to 3 feet (90 cm) long.

Finding Food

A spiny lobster hides among rocks on the ocean floor during the day and comes out at night to feed. It can walk on its legs or paddle along by moving its chunky tail. It uses its large, spiny antennae for fighting and defense. It uses its two smaller antennae to find food. The lobster sweeps the water with these feelers to detect movement and to pick up the scent of any nearby prey.

A spiny lobster's spines are sharp enough to rip skin.

Mussels on the Menu

When the lobster moves in for the kill, it grabs a mussel with its front legs and positions the victim near its mouth. Then it bites down with its hard jaws to break off bits of shell and get to the soft body of the mussel inside. The lobster chews its food into tiny pieces that it can digest, using teeth that are inside a stomach near its mouth. The three grinding teeth look a bit like the surface of human molars—the teeth at the back of our mouths.

After they have fed, spiny lobsters shelter in holes in rocks or reefs.

Extra Extreme

A spiny lobster has the perfect escape technique. It can make one of its claws, walking legs, or antennae drop off if a predator grabs it. This allows the lobster to make a getaway and save its life. Later, the body part regrows, so the lobster is complete again.

Sea Lion

The spines that cover a spiny lobster's shell help protect it from some predators, but not all. Sea lions are ocean predators that can be as fierce as the lions they are named after, and they often catch and eat lobsters.

Deep-Sea Divers

Sea lions can dive to depths of almost 1,000 feet (300 m) to search the ocean floor for food. Their streamlined, torpedo-shaped bodies and powerful front flippers help them swim fast, and their back flippers act as rudders so they can change direction. Sea lions are mammals, so they have lungs and need to come to the ocean surface to breathe air. When they dive, their nostrils close up automatically so they can hold their breath and stay underwater for up to 20 minutes at a time.

A sea lion's sharp teeth are vital tools for catching prey like spiny lobsters.

Sea lions can twist and turn quickly in the water as they chase prey. They are real ocean acrobats!

Sea Lion Senses

Light from above does not penetrate the waters in the deep ocean. In these very dark conditions, sea lions use their excellent senses of smell and hearing to find prey to eat. They also use their sensitive whiskers to feel ripples in the water that tell them lobsters and other animals are moving nearby. When a sea lion finds a lobster, it shoots forward quickly and grabs the prey in its sharp teeth, crushing the life out of the victim in an instant.

Extreme!

Seeing in the Dark

Sea lions can see better in dark waters than many animals. They have a reflective layer at the back of each eye that works a bit like a mirror, bouncing any small amounts of light back through the eyes a second time to help the animals see.

Great White Shark

A sea lion can be ferocious—but it is no match for the great white shark. The shark is an apex predator, which means it sits at the top of the food chain. Great white sharks hunt and eat a variety of prey, and very few animals in the ocean dare to approach or attack these terrifying fish.

Sharp Senses

Great white sharks have several senses that help them find food. They have an amazing sense of smell that can detect even a single drop of blood dissolved in 10 billion drops of water—even more than 1 mile (1.6 km) away. Their ears are two small openings behind and above the eyes that can sense even tiny vibrations made by prey moving in the water. In fact, great white sharks can tell where a prey animal is moving up to 800 feet (250 m) away.

Great white sharks are about 20 feet (6 m) long and can leap out of the water to catch prey—even birds in flight!

Shark Attack

When a shark spots a sea lion, it shoots up through the water at high speed, slamming into the victim with great force. The shark then grabs the animal tightly in its mouth as it leaps from the water. Great white sharks are armed with seven rows of sharp, jagged teeth. These teeth are designed to cut flesh and can easily crush and shatter bone.

Great white sharks can smell sea lions from more than 2 miles (3.2 km) away.

Extreme!

Super Sensors

Great white sharks can also use electrical signals to detect and home in on prey. All living things create electrical signals when they breathe or move, however slightly. The sharks have sensors on their heads that pick up these signals to pinpoint prey's location.

Chapter 2

Rain Forest Rivalries

Rain forests cover about one-twelfth of Earth's land, but they are home to more than half the planet's species. The lives of all these organisms are linked together in thousands of different food chains, all starting with trees.

Tapestry of Trees

Most photosynthesis in a rain forest happens in leaves that grow on the highest branches of trees. This canopy can be 100 feet (30 m) or more above the ground. Many small plants such as orchids grow on branches to get a share of the sunlight. Smaller trees grow in the dappled shade beneath the canopy. Rain forests occupy places with high rainfall, so trees need only shallow roots to get the water they require. The rain forest soil is packed with decomposers that break down the large quantities of dead leaves and animals, which releases nutrients that the trees use to grow and remain healthy.

To avoid toppling over, rain forest trees grow supporting buttress roots from the base of their trunks.

Space Invader

When a tree falls, or there is a forest fire or other disturbance in the rain forest, cecropia trees are some of the first to take over the new space. They grow tall quickly in sunlit gaps, and most of their large leaves are on the upper branches. Cecropia trees are common in rain forests but not in places where the light is blocked by a canopy of taller trees.

Extra Extreme

Cecropia trees defend themselves against leaf munchers such as leafcutter ants with the help of tough little Azteca ants. The trees have hollow stems where the Aztecas can nest, and their leaf bases make tiny white blobs of sugar-rich food the ants like to eat. In return, the Aztecas drive off insects as big as caterpillars, and even take on sloths.

Sloth

Sloths are slow-moving mammals that live in rain forest canopies. They like to eat leaves of trees such as cecropia, and they have very healthy appetites—about two-thirds of a well-fed sloth's body weight is in its stomach.

Slow Food

Sloths usually feed upside down, slowly munching on leaves with their peg-shaped teeth. Rain forest leaves are often tough to digest and may contain few nutrients, so sloths need to eat plenty to get enough goodness and energy from them. Sloths have several stomach compartments, which contain bacteria that help digest the leaves. Even so, digestion takes as long as one month. They supplement their leaf-rich diet with occasional insects, lizards, and even dead animals. Sloths have adapted to their low-energy diet by moving slowly and by maintaining a lower body temperature than many other mammals.

Sloths have either two or three long, curved claws on the toes of each foot that can hook over branches so they can easily reach leaves.

Fur Protection

Most mammals have fur that grows down toward their feet, but a sloth's long brown fur grows in the opposite direction. Since the sloth usually hangs upside down, this helps rain run off its body so it does not get waterlogged. The fur also has tiny plantlike bacteria growing in it. These make many sloths look green and help them blend into their background of leaves and branches.

This sloth has a green color because of the bacteria on its fur.

Extreme!

Danger Zone!

Sloths have amazing self-control. They pass feces and urine just once a week. When sloths do their business, it is a time of extreme danger because they have to leave the canopy, slowly descending tree trunks to the ground. There they may become an easy meal for jaguars or other rain forest predators such as swooping harpy eagles.

Harpy Eagle

The harpy eagle is a giant among birds of prey. It has talons the length of bear claws to grab rain forest prey such as unsuspecting sloths.

Aerial Attack

Harpy eagles rest on high perches in the rain forest, keeping a lookout for meals. When an eagle spots prey, such as a monkey, sloth, or iguana, it attacks at 50 miles per hour (80 kph), using feet the size of human hands to grip and crush the victim with immense power. It then often carries the meal to a high branch, where it gets to work with its sharp, hooked beak, ripping off pieces of flesh to swallow.

The harpy eagle's large, forward-facing, close-set eyes make it easy to spot prey.

Harpy eagles can even catch and eat large howler monkeys like this one.

Canopy Kingdom

A pair of harpy eagles has young every two to three years. The couple make a nest high in the rain forest canopy in the crook of a large branch. The female lays two eggs, but only the first to hatch survives. She delivers pieces of flesh from her catches to the chick throughout the day for up to 10 months. After this the youngster must set up its own territory, big enough for it to find sufficient food to survive and to breed in the future. Harpy eagles need large territories with plenty of food, so they usually live about 10 miles (16 km) apart.

Extra Extreme

The wingspan of the harpy eagle can reach 6.5 feet (2 m). It needs that power to lift its bulky body weight of up to 20 pounds (9 kg). The even heavier Andean condor can weigh 33 pounds (15 kg) and has a wingspan of up to 10.5 feet (3.2 m), the widest of any bird of prey. It uses these huge but narrow wings to glide on mountain updrafts searching for prey.

Chapter 3

Death in the Desert

Deserts are places that get little rain and where it can be burning hot during the day and freezing at night. Fewer plants grow there than in many other habitats, and the animals that live there face a constant struggle to survive.

Surviving in the Sahara

The Sahara Desert in North Africa is the world's largest hot desert. It covers an area of land similar in size to the United States. Many food chains in this desert begin with the tough grasses, shrubs, and trees that grow there. These plants produce their own food by photosynthesis. The insects, such as desert crickets, and other animals that eat the leaves, seeds, and other parts of these plants get both water and energy from this food.

The Sahara Desert is a harsh place for animals to live, so the race is on to eat or be eaten!

A desert cricket's coloring helps camouflage it against the sand and fool predators.

Hungry Crickets

Desert crickets are insects that have long back legs for leaping and can fly using their two long, thin back wings. They travel across the desert in search of leaves and seeds to eat. They are colored green or brown to camouflage them against plants while they eat, so that predators cannot see them. The only time crickets draw attention to themselves is during breeding time, when males rub the jagged edges of their front limbs together to make a musical chirping sound that attracts females.

Extra Extreme

Locusts are close relatives of crickets—but these desert exterminators are truly extreme. Locusts often gather in huge groups of millions of insects, called swarms, which can cover an area of more than 450 square miles (1,200 sq km). When a locust swarm feeds, it destroys everything in its path, including trees, crops, and other plants.

jerboa

Jerboas are long-tailed desert rodents that eat seeds, leaves, and insects such as crickets. Amazingly, these little creatures never drink. They get all the water they need in the bone-dry desert from their food.

Hiding Away

Jerboas stay out of the hot Saharan sun during the day by hiding in burrows that they dig using their teeth, noses, and claws. They search for food at night. A jerboa's back legs are about four times longer than its front legs, and it gets around by leaping. When the little rodent springs off the ground, its long tail helps it balance. The tail also helps prop up the jerboa when it is standing.

Although jerboas are small creatures with delicate legs, they are superb jumpers.

Feeding Time

Jerboas can travel long distances as they hop around the desert in search of something to eat. Their good sense of smell helps them find plants, and their hearing helps them locate insect prey. Like all rodents, jerboas have strong front teeth to crunch through the tough external skeletons of insects. Also, when deserts are very dry, these teeth can gnaw through and chew plant roots under the soil to get the food and water stored there.

Thanks to its long back legs and hopping power, the jerboa can cover up to 6 miles (10 km) a day.

Extreme!

Jumping Jerboas

A jerboa is only about as big as a tennis ball, but it can jump several feet both up in the air or across the land. It jumps quickly from side to side in a zigzag pattern to help confuse the predators that try to eat it, such as owls, snakes, foxes, and jackals.

Fennec Fox

The fennec fox is the last link in our desert food chain. This swift, smart carnivore catches and eats a variety of desert animals, including the fast-hopping jerboa.

Big Ears

The most striking feature of the fennec fox is its huge ears, which can measure up to 6 inches (15 cm) long. Like the jerboa, the fennec fox usually escapes the heat of the desert sun during the day by resting in an underground burrow. If the fox does appear in daylight, its big ears help it lose body heat and keep cool. As blood circulates close to the wide surface of the ears, it is cooled, and this stops the fox from getting too hot.

The fennec fox's long, sandy-colored fur helps camouflage it in the desert. The thick fur also keeps it warm at night.

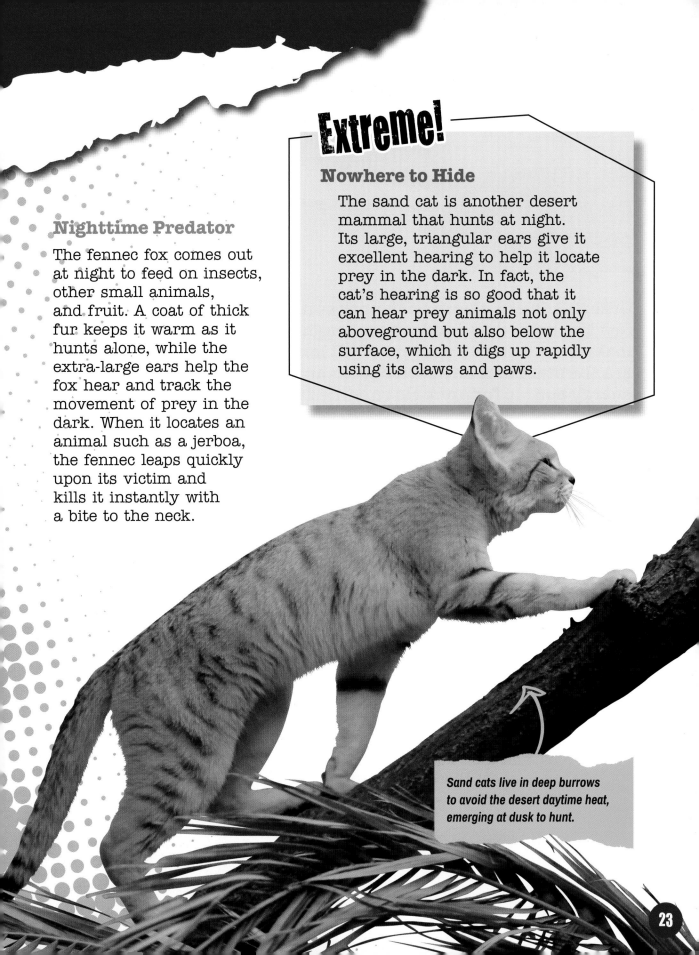

Nighttime Predator

The fennec fox comes out at night to feed on insects, other small animals, and fruit. A coat of thick fur keeps it warm as it hunts alone, while the extra-large ears help the fox hear and track the movement of prey in the dark. When it locates an animal such as a jerboa, the fennec leaps quickly upon its victim and kills it instantly with a bite to the neck.

Extreme!

Nowhere to Hide

The sand cat is another desert mammal that hunts at night. Its large, triangular ears give it excellent hearing to help it locate prey in the dark. In fact, the cat's hearing is so good that it can hear prey animals not only aboveground but also below the surface, which it digs up rapidly using its claws and paws.

Sand cats live in deep burrows to avoid the desert daytime heat, emerging at dusk to hunt.

Chapter 4

Grassland Battle

Grasslands are places with poor soil or dry conditions, where few trees can grow and where tough grasses dominate. They make up around 40 percent of all land in the world. Grasses are the start of many food chains and provide an important energy source for animals ranging from insects to rhinos.

Different Grasslands

Some grasslands are found in places with cold winters. These include prairies in North America and pampas in South America. Other grasslands, called savannas, are in warmer, tropical places with a rainy season. These have more trees and shrubs growing on them than the prairies do. Each type of grassland has particular food chains. In the African savanna, for example, zebras eat grasses and giraffes eat acacia trees, and then lions and hyenas may eat those zebras and giraffes.

Grasslands are home to many animals, including small insects called termites that build giant nests there.

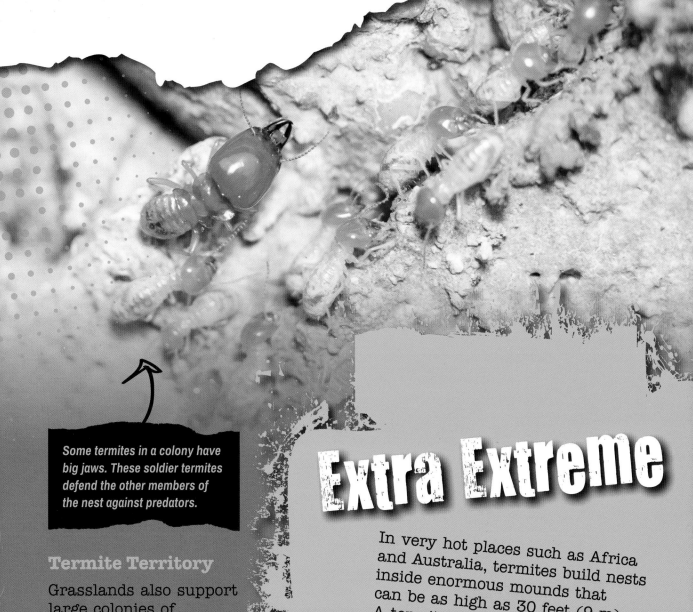

Some termites in a colony have big jaws. These soldier termites defend the other members of the nest against predators.

Extra Extreme

Termite Territory

Grasslands also support large colonies of termites. These small insects build huge nests where a large female— the queen—spends her life laying eggs. Most members of a colony are workers that go out collecting grass, seeds, and other plant parts to bring back to the nest to feed the young. Termites, sometimes called white ants, are similar in size and lifestyle to many types of ants but are actually related to cockroaches.

In very hot places such as Africa and Australia, termites build nests inside enormous mounds that can be as high as 30 feet (9 m). A termite builds a mound from soil, saliva, and dung, with airholes that lead to a central chimney. Air moves through the chimney, keeping the nest at the base of the mound cool. Termites need low temperatures to grow fungi that can decompose wood into food for the insects to eat.

Long-Nosed Armadillo

It may look like a cross between a pig and a tortoise, but the long-nosed armadillo is related to neither of those animals. Instead, this strange creature is a small, armored mammal of the pampas grasslands that likes nothing better than to munch on termites.

Termite Terminator

The long-nosed armadillo snuffles along at dusk or nighttime in search of food, making the most of its good sense of smell to find ants, termites, or other insects to eat. The armadillo uses its large middle toenails to dig into the ground or to rip pieces from termite mounds, and then laps up the insects, dozens at a time, with its sticky tongue. It can also hold its breath for up to six minutes, which is helpful when feeding and digging.

Armadillos hunt for insects that live above and below the ground.

An armadillo uses its front claws not only to uncover food but also to make protective burrows.

Defense Moves

The armadillo has two tactics for evading capture by grassland carnivores such as pumas and hawks. The first is to scuttle off to hide in one of its burrows, which can be up to 20 feet (6 m) long. The other tactic is to duck and cover. An armadillo has thick, bony plates on its forehead and body, forming a rounded shell, with the piece over the back divided into bands separated by seven strips of flexible skin. These plates allow the armadillo to tuck in its head and curl into a protective ball.

Extra Extreme

In their burrows, female long-nosed armadillos give birth to clones. A female produces a single egg, which then, after fertilization, divides into between 8 and 12 identical babies. So, an armadillo litter will always be all males or all females.

Puma

Puma, mountain lion, and cougar are all different names for the same large, slender, powerful cat that hunts in the rocky grasslands of North America and South America. Like the great white shark or fennec fox, the puma is an apex predator—a killer that is rarely the prey.

Stalk and Destroy

Pumas have adapted to find and kill prey. They prefer large prey such as deer but will eat anything from armadillos to snails, even attacking livestock such as poultry and calves. A puma stalks its prey and then leaps at close range, springing from its long back legs. It then uses massive teeth to break the animal's neck. A puma may eat small prey on the spot, but it drags large prey to a hiding place. Then it can return each night to feed.

Pumas see very well during the day and at night, have fine-tuned hearing, and a powerful sense of smell.

The puma's powerful body, long, agile limbs, and sharp teeth and claws make it a formidable hunter. It can even swim and climb trees.

Head Hunter

The puma is at the top of many pampas food chains. Only young pumas, which might be killed by hungry bears or wolves, are ever in danger. The puma is bad news for the individual animals that become its meals, but good news for animal populations in general. Apex predators like pumas usually prey on weak, sick animals that cannot evade them so easily. In this way, they help keep the overall populations healthy.

Extreme!

Lone Hunter

Even though pumas can grow up to 8 feet (2.5 m) long, they are not classified as big cats because they do not roar and are closely related to smaller cats such as the ones we keep as pets. Like tigers and leopards, pumas live and hunt alone. To find enough prey to survive, they need to cover wide areas. Never more than five pumas are found in every 40 square miles (100 sq km).

Chapter 5
River Wipeout

Many extreme freshwater food chains are found in rivers. The water plants that grow there provide food and shelter for many animals, from worms to fish and ducks. The next link in a river food chain, such as a caiman, will survive only if it can eat such prey. In turn, the caiman itself becomes the prey.

Firm-Rooted Plants

The fast-moving waters in a mountain river could damage plant leaves. So plants that grow there, such as eelgrass, have tough roots and long, thin leaves that do not drag in the water. In a wide, deep, slow-moving river, some plants such as water lilies are rooted in the depths but have wide, air-filled leaves. They float at the surface making food by photosynthesis. Other river plants, from reeds to alder trees, are rooted in the mud of riverbanks.

Insect larvae that eat river plants are then eaten by other animals.

Prey for Piranhas

The piranha lives in rivers in South America. Although not a huge fish, it can measure up to 1 foot (30 cm) long and weigh up to 8 pounds (3.5 kg). It is known as a fearsome hunter that uses its razor-sharp teeth to wipe out anything in the water. In reality, piranhas are omnivores that feed mainly on fish, shrimp, and water snails, along with river plants. However, on rare occasions, schools of hungry or provoked piranhas will all feed at the same time on large prey. Then, they can strip a carcass in minutes.

Piranhas are fast swimmers that hunt using a good sense of smell, fast swimming, and triangular teeth that lock together to bite through flesh.

Extra Extreme

Piranhas have extremely powerful bites. Around 2 percent of their whole weight is made up of jaw muscles, and the jawbones lever together to create an immense force. Scientists have estimated that, pound for pound, piranhas can bite with three times more force than alligators.

Caiman

Piranhas can be ferocious fish, but caimans are far more dangerous river inhabitants. These sharp-toothed carnivores, related to alligators and crocodiles, normally reach adult lengths of up to 7 feet (2 m).

Universal Hunter

Caimans hunt anything from water snails and fish such as piranhas to deer. Sometimes they lie on the riverbank to bask in the sun, although usually they are in the water keeping a lookout for food with just their eyes peeking above the surface. Smaller caimans may chase prey underwater, but larger ones lurk at the water's edge. They suddenly lunge from the river to catch large animals coming for a drink and then drag the prey underwater to drown them before dining.

Caimans will wait patiently, with just their eyes above the water, until victims come close enough for them to attack.

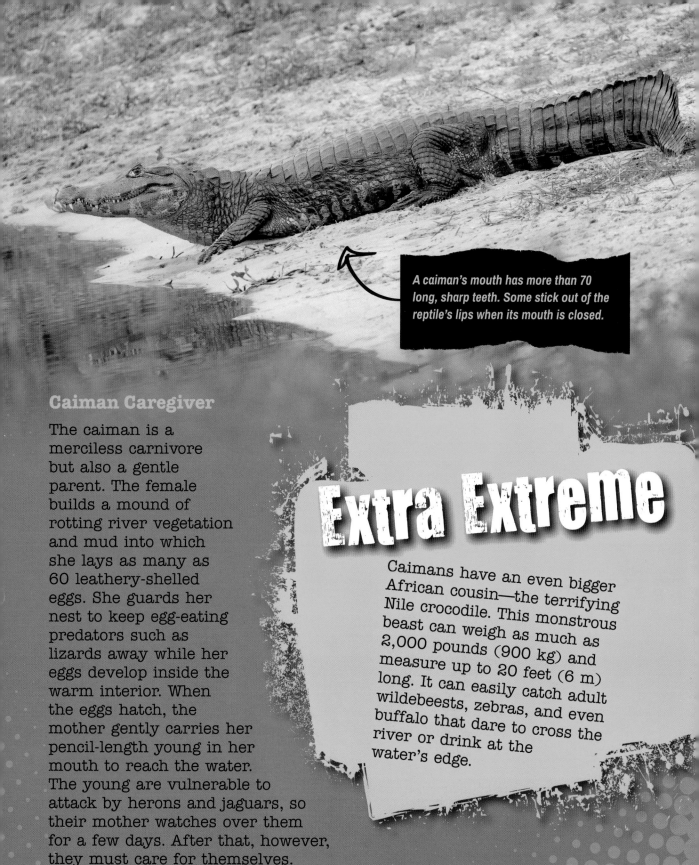

A caiman's mouth has more than 70 long, sharp teeth. Some stick out of the reptile's lips when its mouth is closed.

Caiman Caregiver

The caiman is a merciless carnivore but also a gentle parent. The female builds a mound of rotting river vegetation and mud into which she lays as many as 60 leathery-shelled eggs. She guards her nest to keep egg-eating predators such as lizards away while her eggs develop inside the warm interior. When the eggs hatch, the mother gently carries her pencil-length young in her mouth to reach the water. The young are vulnerable to attack by herons and jaguars, so their mother watches over them for a few days. After that, however, they must care for themselves.

Extra Extreme

Caimans have an even bigger African cousin—the terrifying Nile crocodile. This monstrous beast can weigh as much as 2,000 pounds (900 kg) and measure up to 20 feet (6 m) long. It can easily catch adult wildebeests, zebras, and even buffalo that dare to cross the river or drink at the water's edge.

Jaguar

The dappled coat of the jaguar, with its different shapes and sizes of spots, blends into the riverbank and forest where this big cat lives. It hunts from hiding places in trees and on the ground, but will break cover and leap into the water to hunt prey when it spots an opportunity for a meal.

Skull Piercer

Jaguars hunt for more than 12 hours each day, often at dawn or dusk when many prey cannot spot them approaching. On land, the cats may hunt wild pigs or huge rodents called capybaras, but caimans and turtles in the river are also regular prey. Jaguars tend to ambush prey and grab hold of them in their large, clawed paws. Using the great biting power generated in its jaws, the jaguar then delivers the killer blow: a piercing bite through the skull into the unfortunate victim's brain.

A jaguar is a deadly hunter both on land and in water. Once a jaguar bites into its prey, there is no escape.

Feline Powerhouse

The jaguar is the third-largest big cat. It is built like a pit bull dog, with short but immensely powerful legs and a stocky body that can weigh as much as 250 pounds (113 kg). With its strong claws, a jaguar can easily haul itself up even vertical tree trunks. Unlike most cats, jaguars like to swim, bathe, and even play in rivers and pools.

Extreme!

Night Vision

Jaguars, like other cats, have no problem spotting prey, even in the dark. They have large pupils in their eyes that allow as much light as possible to enter. Also, special membranes across the eyes act like mirrors to reflect any available light onto the retinas at the back of the eyes.

A jaguar uses its sharp teeth to kill prey and to bite, hold, and tear food apart. Its teeth are even powerful enough to crush bone.

Chapter 6
Arctic Attacks

The Arctic is an extreme environment. In winter it is in total darkness, there are blizzard winds, and it is cold enough for seawater to freeze over. There are a few Arctic land plants, but they are mostly stunted, tough bushes. In fact, most photosynthesis in the Arctic happens in the ocean.

Nutrient Cycle

The phytoplankton of the Arctic Ocean thrives under thin ice, which acts like a sunscreen for harmful light during photosynthesis. The tiny plantlike organisms grow using food and nutrients in the water and are eaten by animal plankton, which, in turn, become food for small fish, shrimp, and other animals. Dead plankton and animals sink, refreshing food supplies for anemones, crabs, and other creatures of the Arctic seafloor. Nutrients released by decomposers on the seafloor circulate up to surface waters where they are used by phytoplankton.

The king crab has eyes on short stalks to help it spot food on the floor of the Arctic Ocean.

King of the Ocean

King crab larvae are swimming types of animal plankton. When they grow big enough, they fall 100 feet (30 m) or more to the seafloor and change into non-swimming crabs. Adult king crabs are whoppers of the crustacean world. Large males can weigh more than 20 pounds (9 kg) and stretch 5 feet (1.5 m) across. The crabs feed on animals such as clams and worms, using their long claws to grab food and put it in their mouths.

Extreme!

Arctic Wanderer

King crabs migrate from shallow waters, where they breed and their eggs hatch, to deeper waters where they usually live and feed. This journey can be a 100-mile (160-km) round trip during which crabs may cover 1 mile (1.6 km) a day, which is good progress in such dark, cold water.

Bearded Seal

Hundreds of long, white, sensitive whiskers formed into a drooping moustache give the bearded seal its name. Without those whiskers, the seal would struggle to be part of Arctic food chains.

Hunting in Darkness

A bearded seal takes a deep breath before diving. Even a youngster can dive down 250 feet (75 m) and back, which takes around five minutes. In the dark depths, the seal uses its whiskers to feel prey, including smaller king crabs and shellfish, near its mouth. It also has sensitive hearing to pick up noises made by seafloor animals and large eyes to see moving prey such as Arctic cod in the dark water.

A bearded seal has thick, waterproof fur and a layer of fat called blubber under its skin to keep warm in the Arctic.

Extra Extreme

Arctic Survival

Bearded seals need to return to the surface of the sea to breathe and to rest. As they dive under sheets of ice that have few openings, they dig breathing holes, which they use regularly to make sure the water does not freeze over. Out of the water, the seals' sight is not strong and they also move slowly, making them vulnerable to lurking polar bears. Female bearded seals are especially at risk when they come to land to give birth, but their pups can swim within hours of being born. However, the youngsters are far from safe in the water, as even adult seals cannot always outswim killer whales, or orcas.

The free-diving specialists of the seal family are the Weddell seals of the Antarctic. They regularly descend to feed on the seafloor 2,000 feet (610 m) under the Antarctic ice. However, their larger relatives, elephant seals, have the deepest dive record at more than 4,900 feet (1,500 m). That is 1 mile (1.6 km) underwater.

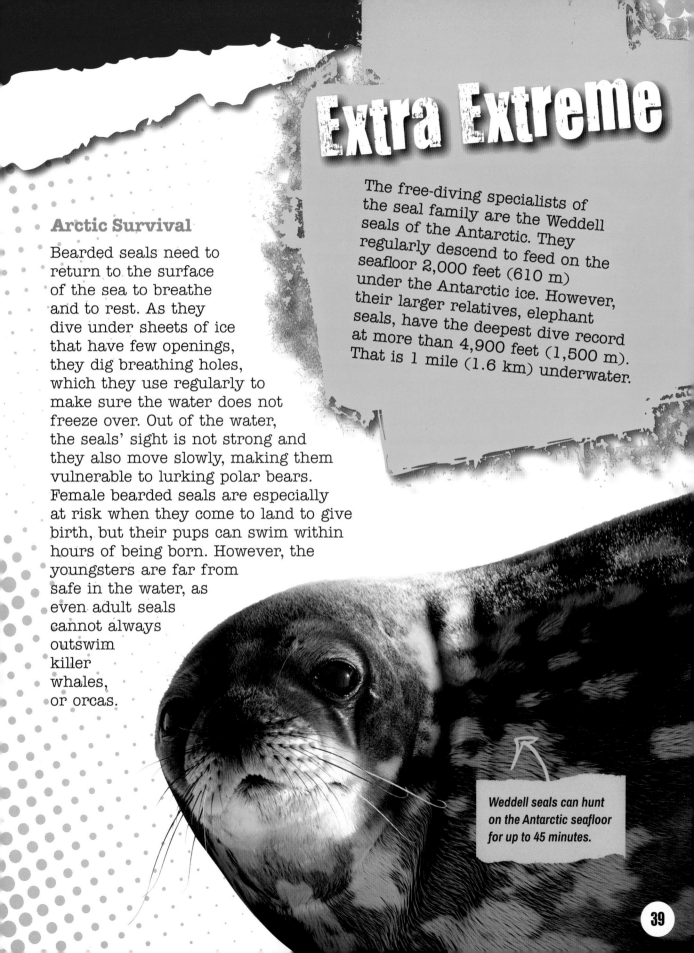

Weddell seals can hunt on the Antarctic seafloor for up to 45 minutes.

Polar Bear

Polar bears are mighty predators of the Arctic Ocean. They hunt for marine prey such as bearded seals on coastal land and on ice sheets lying across the water.

Patient Killer

A polar bear may lie in wait for hours before a seal pops its head out of a breathing hole. It will then try to strike and catch the seal with its paws, drag it from the water, and devour the meal. Polar bears also sniff out seal pups born in dens under the ice. Then they crash through the roofs to capture tasty snacks. Polar bears also eat walruses and dead whales. When prey is in short supply, a bear will even eat birds' eggs and trash.

Polar bears can sniff newborn seal pups from 1 mile (1.6 km) away and under 3 feet (1 m) of snow.

Marathon Meal Break

Female polar bears that have fattened up from eating seals become pregnant in fall. They dig dens in the snow where they have their cubs, which would freeze to death in the Arctic winter without this protection. The female gives birth usually to two cubs in midwinter, when blizzards are raging outside. She feeds them on her rich milk and then emerges in late spring when it is warmer and the cubs are bigger and stronger. She can then start to feed again for the first time in up to eight months.

Polar bears are the largest land predators in the world. Adults can weigh up to 1,700 pounds (770 kg).

Extreme!

Built for the Conditions

As well as fat and dense fur to stay warm, polar bears have other adaptations for life in the Arctic. They have hollow outer fur that reflects light, so they appear white for camouflage in the snow. They have huge paws with webbed toes for swimming, pointed claws that dig into ice, and non-slip soles to prevent them skidding on ice.

Killer Whale

When a polar bear is swimming between large sections of ice, called ice floes, it needs to watch out for killer whales. Although they are called whales, killer whales are actually large dolphins. These apex predators eat even polar bears.

Terminator Teams

Killer whales are intelligent and work as teams, called pods, to catch prey. They ambush larger whales by attacking fast and biting them from different sides. They also work together to make waves that wash resting seals from ice floes into the water, so they can catch them. For an easier meal, killer whales round up schools of fish and stun them with tail slaps. They also hunt alone and target anything from salmon to sharks and dolphins. Blubber-rich prey is what an orca craves, but a polar bear is a rare meal because it is too furry and bony.

No two killer whales have the same black-and-white pattern. The height and shape of the back fin also varies.

Killer whales live and hunt in pods that cruise the Arctic waters in search of prey. They are at the top of Arctic food chains.

Loud Seeker

Killer whales are noisy. They scream, whistle, and make other sounds to communicate with other orcas. A whale also makes a rapid clicking noise from a fatty lump inside its head. These clicks travel through the water and bounce off any prey nearby. The whale detects the echoes through its jaw and uses this information to figure out the location of the prey. This prey-detection system is called echolocation.

Extra Extreme

The biggest toothed whale is the sperm whale. It can reach a length of 60 feet (18 m) and has around 50 teeth in its jaws. Each tooth can measure 8 inches (20 cm) long and weigh as much as 2 pounds (1 kg). The sperm whale uses these teeth to grab slippery giant squids more than 1 mile (1.6 km) underwater.

Chapter 7
Links in the Chain

All links in every food chain depend on each other. Without enough sunlight, grass dies. Without enough grass, rabbits might die, and then foxes may go hungry.

Final Connections

The last link in any food chain, after an animal dies, is a decomposer. Decomposers include tiny bacteria in soil, air, or water, and also fungi, such as mushrooms, on land. These break down organic waste into the nutrients it is made from, which producers at the start of the chain, such as plants, absorb. Decomposers rely on the helping hand of scavengers, including animals such as maggots, wood lice, worms, and vultures, which, as they feed, break down remains into smaller pieces that decomposers can attack.

Fungi grow on and break up dead tree trunks. This clears space for new plants and returns nutrients to the soil.

Broken Links

Many food chains, including some in this book, are under threat globally. For example, rising global temperatures mean there is less ice cover on the Arctic Ocean each year and so less access to seals for polar bears. Deforestation in rain forests means fewer leaves for sloths to eat and therefore fewer meals for harpy eagles. People need to care for the different ecosystems of our planet to ensure that the food chains in them can survive.

Vultures look mean but have a useful job. By clearing away dead, rotting bodies, they help stop diseases from spreading.

Extreme!

Circle of Life

Every ecosystem on Earth has hundreds of food chains that are interconnected in complex food webs. For example, mice in a prairie might eat grasshoppers and grasses, and be eaten themselves by many predators, including coyotes, owls, and snakes. In any food chain or web, there are far more producers than apex predators. Scientists estimate that 1 million phytoplankton and other producers in the ocean might support 10,000 plant eaters, which themselves support just 100 tuna and 1 great white shark!

Glossary

adaptations changes in an animal that help it stay alive

antennae (singular is antenna) the long, thin body parts on an animal's head used to feel things

apex predator a predator at the top of a food chain, with no or few predators of its own

bacteria single-celled living things, some of which can be harmful

bask to lie in a warm place

blubber a thick, tough fat layer found on sea mammals

breed to reproduce or have young

camouflage a color or pattern that matches the surroundings

carnivore a meat-eating animal

clones identical organisms with the same genes as other organisms

colony a group of plants or animals living in the same place

crustacean an animal with a hard shell and several pairs of jointed legs that usually lives in water

decomposers living things that break down dead organisms

digest to break down food inside the body to get nutrients from it

echoes reflected sound signals

echolocation using reflected sound signals to locate objects

ecosystem a group of organisms and where they live

feces solid animal waste

food chains groups of living things, in which each relies on the next in the chain for food

food web a group of connected food chains

fungi (singular is fungus) a group of organisms that includes yeasts, mushrooms, and molds

larvae (singular is larva) young animals, particularly insects, with different forms than adults

mammals warm-blooded animals that have hair or fur and feed milk to their young

nutrients the substances that nourish living things

organic made by a living thing

organisms living things

pampas large areas of grass-covered plains in South America

photosynthesis the process by which a plant makes its own food

phytoplankton tiny floating organisms that can make their own food

prairies extensive, flat lands that are covered mostly in grasses

predator an animal that catches and eats other animals

prey an animal eaten by another animal

producers living things that make their own food

pupils the rings of muscle at the front of eyes

retinas the layers at the back of eyes that convert light patterns into signals to the brain

species a type of living thing

For More Information

Books

Seidensticker, John, and Susan Lumpkin. *Predators* (Insiders). New York, NY: Simon & Schuster Books for Young Readers, 2008.

Solway, Andrew. *Food Chains and Webs* (The Web of Life). North Mankato, MN: Raintree Freestyle, 2012.

Weidner Zoehfeld, Kathleen. *Secrets of the Garden: Food Chains and the Food Web in Our Backyard.* New York, NY: Dragonfly, 2014.

Websites

Try building a food chain at:
teacher.scholastic.com/activities/explorer/ecosystems/be_an_explorer/map/foodweb_play.htm

Learn about the animals of the rain forest by visiting:
kids.mongabay.com/elementary/201.html

Discover the differences between the world's biomes and major ecosystems at:
www.ucmp.berkeley.edu/glossary/gloss5/biome/index.html

Compare the different effects of changing climate on Arctic animals at:
climatekids.nasa.gov/arctic-animals

Index